The Foundations of Science

EARTH

Exploring Our Home

WORKBOOK

TAN Books
Gastonia, North Carolina

Cover & interior design and typesetting by www.davidferrisdesign.com

ISBN: 978-1-5051-1903-9

Published in the United States by
TAN Books
PO Box 269
Gastonia, NC 28053

www.TANBooks.com

Printed in the United States of America

EARTH

Exploring Our Home

“Where were you when I laid the foundation of the earth?”

–Job 38:4

CONTENTS

A NOTE TO PARENTS

Thank you for using *The Foundations of Science* series to educate your child about God's wonder-filled world of science! Before diving in, make sure to read these brief notes.

WORKBOOKS' PURPOSE

The workbooks in *The Foundations of Science* series are meant to be a companion to the texts, a simple tool you can use to ensure your child comprehended the material. But it's also supposed to be fun! The exercises should not feel like a test. Consider letting them use the text as they answer questions since we just want them to understand the main concepts and remember some of the things they have learned. (We are not trying to stump them!) Younger students especially may need a little "hand-holding" to get some of the answers, but that's okay, and is even encouraged.

TARGET AGE

The workbook is perfect for middle elementary-aged students, but children as young as first grade or as old as fifth can engage with it. There are enough activities that not every one should be done, and the age of the child can be used to determine which are completed. For example, coloring pages can be used for younger students, while older children may skip those; conversely, younger students may skip some of the personal reflection short answers, while the older ones may be expected to not only answer them but write good and complete sentences. Please cater the workbook to your family's needs.

TYPES OF ACTIVITIES

Most chapters will utilize both a substantive activity (Matching, True/False, Short Answer, etc.), along with something fun, such as a puzzle, word search, coloring page, or arts and crafts. There are also some personal reflection exercises, and most chapters include a question or activity that ties what they studied in that chapter back to the Catholic faith.

MY SCIENCE JOURNAL

Every chapter begins with a "My Science Journal" spread. Here the students are encouraged to take notes as they read the text, write down questions they have, and list the most interesting thing they learned in that chapter, or what they enjoyed the most. They can also log things they saw in nature that week (and it does *not* have to be things that relate to that week's content). This is highly recommended to complete, as it not only helps them comprehend the content better but allows parents to assign a writing exercise as well.

ANSWER KEY

While many of the exercises are subjective and answers will vary, there are also plenty of objective answer exercises that will require grading. An answer key is provided in the back of the book for your use and convenience. If you like, have a conversation about honesty and integrity with your child as you teach them to not peek in the back.

KEY TERMS AND AMAZING FACTS ABOUT EARTH

The Key Terms and Amazing Facts About Earth included in the text are included here as well. Consider making flash cards for the terms to test students' knowledge and retention, and let your children sit and relax as they read the facts; seeing them all at once, rather than buried in the text, may help them remember all the fun things they have learned.

WE ARE HERE TO HELP!

We hope we have provided you with everything you need, but if not, don't hesitate to reach out to your friends at TAN Books with any questions you might have.

CHAPTER 1

SPACESHIP EARTH

MY SCIENCE JOURNAL

NOTES:

MY FAVORITE PART OF THIS CHAPTER WAS:

ONE NEW THING I LEARNED WAS:

SOME QUESTIONS I WANT TO ASK ABOUT THIS CHAPTER ARE:

ONE INTERESTING THING I SAW IN NATURE THIS WEEK WAS:

FILL IN THE BLANK

Use the word bank to complete the sentences correctly.

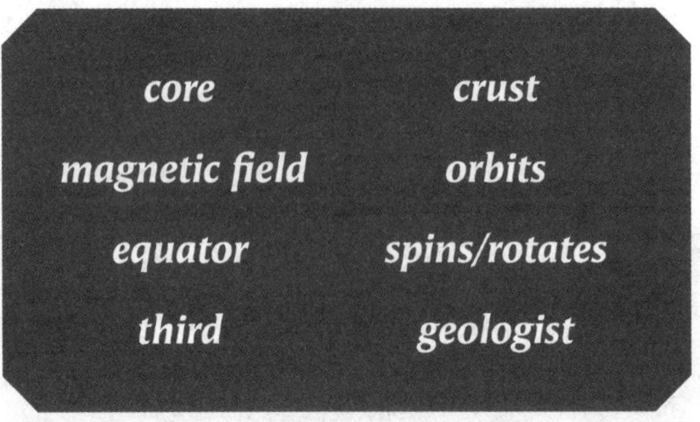

1. The earth is the __________ planet away from the sun.

2. The earth _________ the sun, meaning it circles it in long, oval patterns.

3. Not only does the earth travel around the sun, it also _________ on its axis.

4. The _________ is the imaginary line that circles the earth like a belt, dividing the Northern and Southern Hemispheres.

5. The outer layer of the earth is called the _________.

6. The inner most layer of the earth is called the _________.

7. A __________ is a scientist who studies the rocks and other elements and materials that make up our planet.

8. The earth's ___________ allows us to navigate using a compass.

DAYS AND NIGHTS

In the box below, or in a discussion with your parents, explain why we have day and night.

THE FOUR SEASONS

In the box below, or in a conversation with your parents, explain why we have the four seasons.

COMPASS NAVIGATION

In the box below, or in a conversation with your parents, explain how and why we are able to use a compass to help us navigate. Do outside research if you need to.

MAZE

In the text, we learned that some birds have small pieces of metal in their bills to orient them to the earth's magnetic field, which helps them navigate. That doesn't mean they still don't need some help from time to time. Help this bird find his way back to his nest!

FAITH AND SCIENCE

Color the illustration of the feast of Pentecost, when the Holy Spirit descended like fire on Mary and the apostles as they were gathered in prayer in the Upper Room, then discuss the following questions with your parents:

1. Why is Pentecost considered the birth of the Church?
2. Mary and the apostles were gathered for nine days of prayer in the Upper Room before the Holy Spirit descended upon them. What prayer tradition is derived from these nine days of prayer?

CHAPTER

2

THE EARTH IN MOTION

MY SCIENCE JOURNAL

NOTES:

MY FAVORITE PART OF THIS CHAPTER WAS:

ONE NEW THING I LEARNED WAS:

SOME QUESTIONS I WANT TO ASK ABOUT THIS CHAPTER ARE:

ONE INTERESTING THING I SAW IN NATURE THIS WEEK WAS:

TECTONIC PLATE REVIEW

Answer the questions below.

1. Tectonic plates are:

2. The four ways tectonic plates can interact with each other (four ways they can move) are:

3. Some results of the movement of tectonic plates are:

MOUNTAINS

List at least three facts that you learned about mountains.

1. ______________________________

____

____

2. ______________________________

____

____

3. ______________________________

____

____

FAITH AND SCIENCE

St. Damien Prayer Card

In the space below, draw a picture of St. Damien of Molokai (find an image of him online if you would like). Then, cut it out and write a prayer on the back to him, creating your very own St. Damien prayer card.

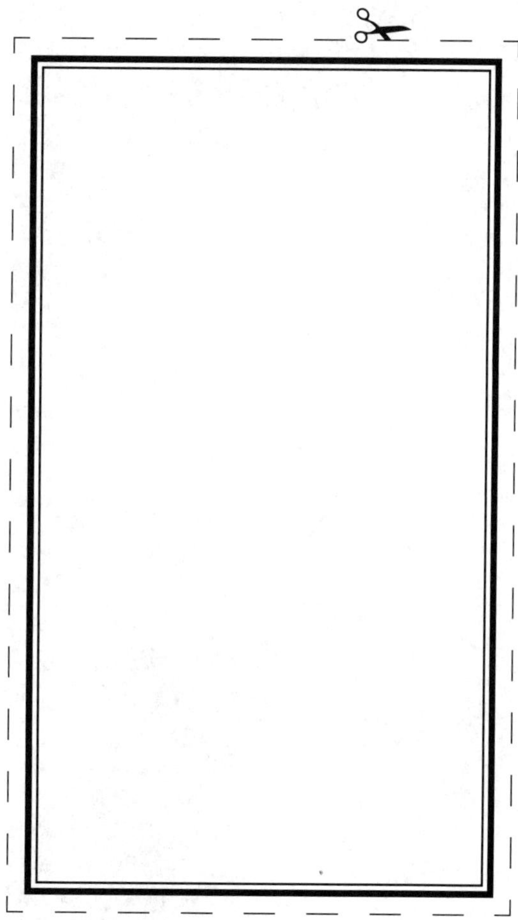

ACTIVITY

Erupting Volcano

We learned about volcanos in this chapter, so let's watch one erupt!

Materials

1. Empty cereal box
2. Tape
3. Scissors
4. Bucket
5. Rectangular foil pan with high sides or old roasting pan
6. Mud (or modeling clay or playdough)
7. Red food coloring
8. 2 Tablespoons Baking Soda
9. White vinegar as needed
10. 2 cups water
11. 2 Tablespoons dish soap
12. Empty 16.9 oz plastic water bottle
13. Measuring cup with a spout
14. Funnel or measuring cup with a spout

Directions

1. Disassemble a large cereal box so that it is broken down and flat. No cutting necessary for this part.
2. With the now flat cereal box in front of you, cut out one of the rectangle-shaped pieces from the cardboard.
3. Roll the rectangle from one corner to the other until you have a cone shape. You will have excess on the bottom which you can trim in step 5.
4. Place the empty water bottle (no lid) inside and shape the cone around it until it fits nicely.
5. Tape it together so that it stays firmly assembled. Trim off any excess cardboard from bottom.
6. Outside, fill the bucket with dirt and water to make a *thick* mud (not too wet).
7. Put some of the mud onto the bottom of the pan.
8. Place the cone into the pan and shape the mud around the cone to form a volcano shape. Eventually, the mud will dry if you don't make it too wet. (Alternate plan: in place of homemade mud, you may use modeling clay or playdough to form the volcano around the cone; feel free to paint the clay brown if it is white.)
9. Mix in 5–7 drops of red food coloring and a couple of drops of the dish soap into the 2 cups of water. Mix lightly together.
10. Pour that mixture into the bottle until it is about 3/4 full.
11. Place the 2 Tablespoons baking soda into the bottle. Use a funnel if necessary.
12. When you are ready to watch the volcano erupt, pour in a little vinegar.
13. Watch the volcano erupt!

CHAPTER

3

SCRATCHING THE SURFACE
Physical Geography and Landforms

MY SCIENCE JOURNAL

NOTES:

MY FAVORITE PART OF THIS CHAPTER WAS:

ONE NEW THING I LEARNED WAS:

SOME QUESTIONS I WANT TO ASK ABOUT THIS CHAPTER ARE:

ONE INTERESTING THING I SAW IN NATURE THIS WEEK WAS:

UNDERSTANDING THE DIFFERENCE

Define the following terms and explain the difference between them.

Weathering and Erosion:

Mechanical Erosion / Weathering and Chemical Erosion / Weathering:

MATCHING

Match up the types of landforms with their definitions below.

A. *Plateau*

B. *Peninsula*

C. *Valley*

D. *Gorge / Canyon*

E. *Island*

F. *Cave*

G. *Basin*

H. *Isthmus*

I. *Buttes / Mesas*

1. ____ A dip or depression in the landscape, sort of like a bowl.

2. _____ Deep, narrow valleys that water carved into rock over the course of many centuries.

3. _____ A piece of land surrounded by water on three sides.

4. ____ Isolated hills with steep, cliff-like sides and a flat top.

5. ____ A narrow strip of land that connects two larger areas of land and separates two bodies of water.

6. ____ A piece of land surrounded by water on all sides.

7. _____ Low-lying areas running between mountains or hills.

8. ____ Underground chambers in the earth.

9. ____ A flat, table-like area of land elevated from the area around it on at least one side.

LATITUDE AND LONGITUDE

First, draw a line to connect each with its definition, then complete the activity.

Equator	0 Degrees Longitude Line
Latitude	Runs up and down and gives measurements for how far east or west a given location is.
Longitude	Runs left to right and gives measurements for how far north or south a given location is.
Prime Meridian	0 Degrees Latitude Line

ACTIVITY

If you do not have a balloon in the house, use a globe to do some research about Latitude or Longitude, finding your hometown and any other major locations you wish to find, or go online with your mom or dad to look at maps.

1. Blow up a balloon of any color. Make it a large as you can (without popping it!).
2. With a sharpie, draw a solid line laterally across the center, like a belt, and label it: Equator.
3. Then draw a vertical solid line in the center and label it: Prime Meridian.
4. Draw two dotted (latitude) lines going horizontally above the Equator (evenly spread apart) and label them 30 N and 60 N. Do the same below the Equator (30 S, 60 S). At the top of the balloon draw a dot to represent 90 N, for the North Pole, and 90 S for the South Pole.
5. Draw two dotted (longitude) lines going vertically to the left of the Prime Meridian (evenly spread apart) and label them 60 W and 120 W. Do the same to the right of the Prime Meridian (60 E, 120 E). Then draw a vertical line all the way around the balloon that is opposite the Prime Meridian (halfway around the world) and label it 180 degrees (this is called the antemeridian, and is also the International Date Line).
6. Once you have these coordinates drawn and labeled, mark the rough location for Vatican City with a "VC" at 42 N, 12 E.
7. Do research to find the coordinates of your hometown and mark those as well, labeling them "ME".

FAITH AND SCIENCE

In the space provided, or on a scrap sheet of paper, write a short essay about how erosion and weathering can serve as images that help us understand the harmful effects sin has on our souls.

CHAPTER 4

SOMETHING IN THE ATMOSPHERE

MY SCIENCE JOURNAL

NOTES:

MY FAVORITE PART OF THIS CHAPTER WAS:

ONE NEW THING I LEARNED WAS:

SOME QUESTIONS I WANT TO ASK ABOUT THIS CHAPTER ARE:

ONE INTERESTING THING I SAW IN NATURE THIS WEEK WAS:

CROSSWORD PUZZLE

Fill in the crossword based on what you read in this chapter.

ACROSS

4. The second layer of the atmosphere—contains the ozone layer.

5. Clouds with distinct edges; puffy or cotton-like.

7. The first layer of the atmosphere.

8. Clouds that form higher in the sky and look thin and wispy, almost like strands of thread or hair draped across the sky.

DOWN

1. Clouds that are low to the ground; tend to be formless and flat, and hazy in the sky.

2. The outer most layer of the atmosphere.

3. The term used to describe all of the gases (oxygen, carbon dioxide, nitrogen, and others) that surround a planet.

5. A visible mass of condensed water vapor and other small particles floating in the atmosphere, usually high above the ground.

6. Sunblock protects us from this kind of light.

ORDERING THE LAYERS OF THE ATMOSPHERE

The various layers of the earth's atmosphere are listed below. On the lines beside them, write them in order from the lowest to the highest.

Stratosphere	1. ______________________
Thermosphere	2. ______________________
Mesosphere	3. ______________________
Troposphere	4. ______________________
Exosphere	5. ______________________

GREENHOUSE EFFECT

What is the greenhouse effect? Why do we call it this? How is it a good thing but also could be bad if we have too much of it?

ACTIVITY

Cloud in a Bottle Experiment

Want to make your own cloud? Read on to find out how!

Materials

1. Bicycle tube valve cut off from old tube with approximately one square inch tubing left below the valve. (Can be acquired from a bicycle fix-it shop. Just ask them to cut one off from an old tube.)
2. 2 Liter clear plastic soda bottle with cap, labels removed.
3. Rubbing alcohol
4. Bicycle pump
5. Gorilla glue (or similar)
6. Gorilla waterproof tape
7. Utility knife
8. Scissors
9. Nail
10. Hammer
11. Electric drill and 1/4 inch drill bit
12. Safety goggles

Instructions

1. Remove the cap from the bottle, and using a small hammer and nail, start a small hole in the center of the cap.
2. Use the drill with the 1/4 in. drill bit to finish off the hole if necessary. Make sure the hole goes all the way through the cap.
3. Centering the flat, square, rubber flap of the valve onto the hole in the cap, glue it down, being careful not to allow glue to get into the hole you drilled. Allow to dry for several hours (or according to instructions on glue).
4. When glue is dry, cut pieces of tape and tape down the square flap of rubber onto the cap. Also tape all around the cap to form an airtight seal, being careful not to go under or inside the cap, as it will need to be screwed back onto the bottle.
5. Pour about a tablespoon of rubbing alcohol into the bottle.
6. Gently roll it around to coat the entire bottle.
7. Place the cap/valve onto the bottle, tightening it as much as possible.
8. Pump air into the valve for 10–15 pumps or until the pump is getting tight. This means you are getting air into the bottle. (Hint: if you hear air hissing, your cap/valve needs to be adjusted in steps 1–4).
9. Quickly remove the pump from the cap/valve.
10. Immediately and quickly unscrew that cap from the bottle.
11. After a slight pop, a "cloud" should appear in the bottle.
12. If the experiment fails, continually check for leaks in your cap/valve and adjust as needed.

TYPES OF CLOUDS

Draw a line from the name for each type of cloud to the picture that represents it.

Stratus

Cumulus

Cumulonimbus

Cirrus

CHAPTER 5

THE WATER CYCLE

MY SCIENCE JOURNAL

NOTES:

MY FAVORITE PART OF THIS CHAPTER WAS:

ONE NEW THING I LEARNED WAS:

SOME QUESTIONS I WANT TO ASK ABOUT THIS CHAPTER ARE:

ONE INTERESTING THING I SAW IN NATURE THIS WEEK WAS:

THE WATER CYCLE

In the box below, explain what the "water cycle" of earth refers to.

On each of the blanks, fill in the right term for each stage of the water cycle.

This happens once the water vapor enters the atmosphere and, as it rises, it cools down, turning it back into liquid form.

____

This happens when too much water condenses in clouds, meaning more and more water molecules stick together in liquid form. This makes the water droplets heavier, and they could fall back to the earth.

____

This refers to when water is evaporated from plants.

____

This happens when heat energy from the sun warms the earth so that the temperature in bodies of water slowly increase, turning some of the water molecules into gas, or water vapor.

____

This refers to liquid water that flows over or across land surfaces.

____

This refers to when plants bring water into their bodies through their roots and other tissue.

____

In the box below, or in a discussion with your parents, explain how plants can have positive benefits on the water cycle.

THE IMPORTANCE OF WATER

In the space provided, list three ways water is important to life on earth. Then in the box, list a few ways that you have used water in the last week.

1.

2.

3.

ACTIVITY

Edible Water Cycle Display

Follow the directions to make an edible water cycle display! Review the lesson as you make it.

Ingredients

1. Large baking sheet pan with side
2. White cotton candy
3. Candied slice lemons (2 halves)
4. Blue raspberry Jell-O, slightly set
5. Light blue canned frosting
6. Piece of chocolate cake or brownie
7. Coconut, 1–2 cups
8. Green food coloring
9. 3.5 x 5 index card
10. 4 toothpicks
11. Scissors
12. Marker

Directions

1. Prepare the Jell-O, allowing it to slightly set to resemble a loose mixture not quite set.
2. Prepare the chocolate cake. When cool, crumble it to resemble dirt. Set aside.
3. Color the coconut with green food coloring to make the grass. Let dry.
4. Begin assembling the display by frosting the top third of the sheet pan with the light blue frosting to resemble the sky.
5. Fill in the middle third of the pan with brown crumbles for the land.
6. Fill in the bottom third of the pan with the loose Jell-O for the water.
7. Place cotton candy pieces in the sky for the clouds.
8. Place two halves of the candied lemon slices together for the sun in the sky.
9. Place green coconut around parts of the chocolate cake crumbles for the grass.
10. Cut up the index card into 4 equal pieces.
11. Write the words "Condensation," "Precipitation," "Evaporation," and "Collection" on each, taping onto a toothpick.
12. Place the labels of the water cycle in the correct place, drawing arrows on each to show a cycle.

Optional: For more detail, use the cake crumbles and coconut grass to create hills and valleys and the Jell-O to show water run-off collecting in streams or rivers.

CHAPTER 6

WEATHER
A Measure of Today

MY SCIENCE JOURNAL

NOTES:

MY FAVORITE PART OF THIS CHAPTER WAS:

ONE NEW THING I LEARNED WAS:

SOME QUESTIONS I WANT TO ASK ABOUT THIS CHAPTER ARE:

ONE INTERESTING THING I SAW IN NATURE THIS WEEK WAS:

ANSWER THE QUESTIONS, OR DISCUSS WITH YOUR PARENTS.

1. Explain the difference between weather and climate.

2. List several different kinds of precipitation.

3. Define meteorology.

4. Explain how RADAR works. What does it stand for?

5. Why does a clap of thunder follow a bolt of lightning?

WORD SEARCH

Find all the terms in the word search below.

Weather	**Humidity**	**Hail**
Climate	**Rain**	**Wind**
Lightning	**Snow**	
Thunder	**Sleet**	

```
A L P X D F W Z G H S B Y E J
C Q O N Z L E O T J R N T D G
O M I Z I H A Y S M Z A Y O L
T W A O C J T A C L M H E A R
A H O D A N H P V I E P K I Y
W F U W Y C E S L H S E P Z W
K U P N W U R C N F H Y T E S
C I L V D Y N M I H A F T M T
R C F O N E R O M U I M I R J
G F J I D F R N O M L J K D U
D I A J X X R I Y I C N K V S
O R P W Q V L S F D H S N O W
N L Z K S B G I R I L T C G U
L I G H T N I N G T S O C Z I
U I J P Q T R R E Y R V G H M
```

ACTIVITY

Make a Rain Gauge

Follow the directions to make your very own rain gauge.

Materials

1. 2 liter soft drink bottle or milk carton
2. Scissors
3. Sharpie / Marker
4. Ruler
5. White tape or duct tape
6. Pebbles / Sand

Directions

1. Cut the top half of the bottle off (make sure an adult is present).
2. Place pebbles or sand in the bottom of the bottle (this will keep it weighted down).
3. Turn the top part of the bottle upside down and place it in the bottom half, creating a funnel-like effect. You may want to tape it to make it secure.
4. Run a piece of tape up the side of the bottle.
5. Using the ruler, mark off inch marks on the tape.
6. If you need to, pour water into the bottle until it reaches the bottom strip on your scale to ensure you are starting at zero.
7. Place outside in an area that is open to the sky and wait for rain!

INTERVIEW A METEOROLOGIST

Meteorologists are scientists who study the weather. Consider finding one to interview (call a news station or find one online). What questions would you want to ask him or her about the weather? Write them in the space provided, or use your own journal.

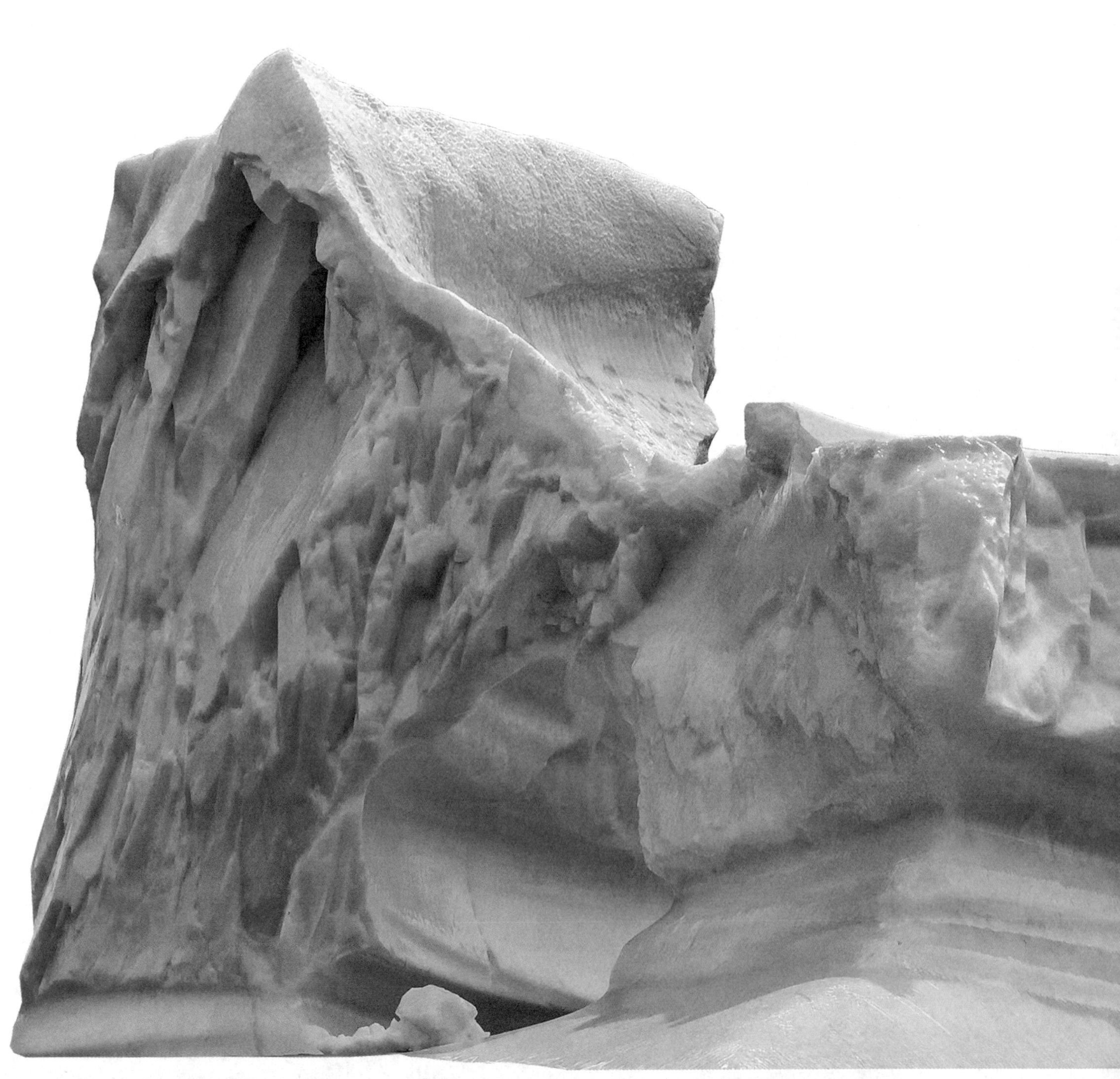

CHAPTER

7

CLIMATE
A Measure Over Time

MY SCIENCE JOURNAL

NOTES:

MY FAVORITE PART OF THIS CHAPTER WAS:

ONE NEW THING I LEARNED WAS:

SOME QUESTIONS I WANT TO ASK ABOUT THIS CHAPTER ARE:

ONE INTERESTING THING I SAW IN NATURE THIS WEEK WAS:

MULTIPLE CHOICE

Choose the best answer.

1. The transfer of heat through air molecules is known as:
   A. transpiration.
   B. convection.
   C. condensation.

2. It is warmer if you live near the:
   A. beach.
   B. Prime Meridian.
   C. Equator.

3. The three major atmospheric circulation cells in each hemisphere break up the globe into three different regions: the tropical zone, the temperate zone, and the:
   A. polar zones.
   B. equator zones.
   C. tundra zones.

4. The climate of a very small region is known as a:
   A. mini-climate.
   B. microclimate.
   C. tiny climate.

5. The effects on a climate caused by being near the ocean are called:
   A. oceanic effects.
   B. coastal effects.
   C. maritime effects.

ACTIVITY

The Climate Near Your Home

Do outside research on the area where you live (city, town, etc.) and answer the following questions about your climate.

1. How much rainfall does your town receive each year?

2. On average, what is the hottest month of the year?

3. What is the highest recorded temperature in that month?

4. On average, what is the coldest month of the year?

5. What is the lowest recorded temperature in that month?

6. How much snow does your town receive each year?

Now pick a location somewhere in the United States (any city of your choosing) that is far away from your home and answer the same questions (draw a "/" after your answers above and record them there). How does this climate differ from yours? Why do you think there are these differences?

continue on next page...

ACTIVITY *(continued)*

To understand the climate of the United States, continue with your meteorology research and answer the following questions about our country.

1. Which state receives the most rain each year?

2. Which state receives the most snow?

3. On average, what is the hottest state in the country?

4. On average, what is the coldest state in the country?

5. Where does your home state fall in these rankings: Most rain or snow, or hottest or coldest?

6. Why do you think you got the answers you did? What is it about these states that make them hot or cold, or cause them to get a lot of snow or rain?

Finally, consider: If you were to investigate the climate in Brazil (or any country in South America), how would it differ? Remember the equator effect!

FAITH AND SCIENCE

Color the picture of St. Kateri Tekakwitha, the patron of the environment. After you are finished, do outside research to write down three facts about her.

1. ______________________________

2. ______________________________

3. ______________________________

CHAPTER 8

NUTRIENT CYCLING
What Goes Around Comes Around

MY SCIENCE JOURNAL

NOTES:

MY FAVORITE PART OF THIS CHAPTER WAS:

ONE NEW THING I LEARNED WAS:

SOME QUESTIONS I WANT TO ASK ABOUT THIS CHAPTER ARE:

ONE INTERESTING THING I SAW IN NATURE THIS WEEK WAS:

WHAT IS THE NUTRIENT CYCLE?

This chapter dealt with the cycle of nutrients through our environment. In the space provided, or in a discussion with your parents, explain what the nutrient cycle is.

UNDERSTANDING THE THREE SPHERES

Put the following spheres of earth in order from inside to outside (low to high) and define what each is:

Biosphere Atmosphere Lithosphere

1.

2.

3.

UNDERSTANDING NITROGEN

Nitrogen was the element discussed most in this chapter. Answer the following questions about this important element.

1. What is helpful about nitrogen?

2. What were some of the ways nitrogen cycles through our environment?

MAZE

Get the nitrogen molecule from the atmosphere to the biosphere!

CHAPTER

9

BIOMES
Unique Habitats and Where to Find Them

MY SCIENCE JOURNAL

NOTES:

MY FAVORITE PART OF THIS CHAPTER WAS:

ONE NEW THING I LEARNED WAS:

SOME QUESTIONS I WANT TO ASK ABOUT THIS CHAPTER ARE:

ONE INTERESTING THING I SAW IN NATURE THIS WEEK WAS:

UNDERSTANDING BIOMES

What is a biome, and what are the main factors that influence it?

NAME THE TERM

What is the term we give to the phenomenon by which a given biome will have different levels of fluctuation (some drastic, some minimal) throughout the year in their temperatures and amounts of precipitation based on where it is located on the earth?

____

TYPES OF BIOMES

Match the four main categories of biome on the left with the description on the right.

Desert	Extremely cold and dry areas, usually found at higher latitudes or elevations.
Tundra	An open prairie dominated by grass species.
Forest	Dry areas with small amount of precipitation.
Grassland	Any area where the dominant plant life is trees.

Label the image of each biome below with the correct biome from the list above.

EXPLORATION ESSAY

In the space provided, write a short essay about one of the four main categories of biomes (desert, forest, grassland, tundra). Pretend you are an explorer and write the essay as if it is a journal entry describing your exploration. To begin, tell us where you are (what continent/country). Then, what sorts of plants and animals would you see? What would the temperature be like? Would it rain or snow? Do outside research to help you.

CHAPTER 10

BIOMES II
Tropical and Temperate Habitats

MY SCIENCE JOURNAL

NOTES:

MY FAVORITE PART OF THIS CHAPTER WAS:

ONE NEW THING I LEARNED WAS:

SOME QUESTIONS I WANT TO ASK ABOUT THIS CHAPTER ARE:

ONE INTERESTING THING I SAW IN NATURE THIS WEEK WAS:

Matching

Match the terms with their definitions.

A. *Deciduous*

B. *Tropical grasslands*

C. *Epiphytes*

D. *Coniferous trees*

E. *Tropical rainforests*

F. *Steppes*

G. *Temperate rainforests*

H. *Biodiverse*

1. ____ Sometimes called "evergreens"

2. ____ Regions where many plant and animal species are found

3. ____ Plants that grow on other plants

4. ____ Also called savannas; known for having rainy and dry seasons

5. ____ Plants that lose their leaves and regrow them each year

6. ____ Also called prairies; often converted into agricultural fields

7. ____ Jungle-like biome; little seasonality, high temperatures, and lots of rainfall

8. ____ Biome with lots of rain (moist conditions) but with a distinct winter season

FAITH AND SCIENCE

Catholic Culture and Traditions Around the World

Similar to the physical biomes found on earth, there are also "cultural biomes"—diverse groups of people who have their own traditions and way of life. This goes for the Church as well. Though we are all one Church—the Body of Christ—the traditions and styles of worship and celebrations may differ from country to country.

By whatever means you wish (pick at random, spin a globe and blindly point your finger, or throw a dart at a map!), select a country other than the United States. The beauty of the Church is that it is universal—found all over the world!

Do outside research about the Church in that country. How many Catholics live there? How did the Church come to this land? What missionaries evangelized it? Who are their most well-known saints? What interesting customs or feast days do they celebrate there?

Put together a report in a separate journal with all the information you want to include. Spruce it up with pictures, drawings, maps, and more. Pretend it is a Catholic tourist brochure you want to hand out to convince people to visit this country.

When you are finished, pray for the people of this country and ask that Jesus and Mary watch over them.

CHAPTER 11

BIOMES III
Extreme Habitats

MY SCIENCE JOURNAL

NOTES:

MY FAVORITE PART OF THIS CHAPTER WAS:

ONE NEW THING I LEARNED WAS:

SOME QUESTIONS I WANT TO ASK ABOUT THIS CHAPTER ARE:

ONE INTERESTING THING I SAW IN NATURE THIS WEEK WAS:

EXTREME HABITATS

Why do we call some habitats extreme, and what are some examples?

FILL IN THE BLANK

Use the word bank to complete the sentences correctly.

Permafrost	*Tundra*
Hibernation	*Boreal*
Chaparral	*Endemic*
Taiga	*Alpine tundra*

1. Boreal forests are also called coniferous forests, or __________.

2. __________ is a long period of sleep-like behavior, which helps animals survive in the winter.

3. Mediterranean scrublands, or __________, have almost no rainfall for several months each year in the summer, but the winters are cooler and wetter. The harsh summer months create drought conditions, and fires often play an important part in the ecosystems found here.

4. __________ is a layer of soil that stays permanently frozen.

5. It is difficult for animals to survive in the __________, but mountain goats, snow leopards, and a few other animal species can be found there.

6. __________ forests have warm, wet summers, followed by bitter cold and drier winters, and are dominated by coniferous trees.

7. __________ are extremely cold biomes found in northern regions.

8. __________ species are animals found in a given region that are found nowhere else on earth.

ANIMALS AND PLANTS WITHIN THEIR BIOMES

Looking back over what you have learned the last three chapters, list the biome where you would expect to find each of the following plants and animals.

Grass

Black Bear

Prairie Dog

Fir Tree

Tiger

Camel

Oak Tree

CHAPTER 12

CONSERVATION
Caring for Our Earthly Home

MY SCIENCE JOURNAL

NOTES:

MY FAVORITE PART OF THIS CHAPTER WAS:

ONE NEW THING I LEARNED WAS:

SOME QUESTIONS I WANT TO ASK ABOUT THIS CHAPTER ARE:

ONE INTERESTING THING I SAW IN NATURE THIS WEEK WAS:

POLLUTION

Define pollution and give a few examples. Why is it harmful to our planet, and what spiritual lessons can we learn from trying to keep our home clean?

BECOME A JUNIOR CONSERVATIONIST!

Think of an activity you can do to help clean up the environment near your home. Can you go on a "litter hike" where you pick up trash in your neighborhood? Can you think of unique ways to use less resources throughout your day? Can you plant a garden and grow your own food? Be creative!

FAITH AND SCIENCE

Thank God for our Home

To complete this unit on Earth, write a letter in the space provided thanking God for our earthly home. Be specific in what you thank him for. What parts of the earth do you enjoy the most, either going there or just learning about? Do you like the beach? The mountains? Certain kinds of animals? Show your gratitude to God by enjoying the wonder that fills the earth!

ZSCHECKEL.

AMAZING FACTS ABOUT EARTH

- If you could burrow like a woodchuck straight through the center of Earth, you'd discover our planet is almost eight thousand miles across. If you traveled around *the outside* of the earth, you would travel almost twenty-five thousand miles to return to the point where you started.
- The earth travels at a speed of around sixty-seven thousand miles per hour as it orbits the sun! But as it moves, it's also rotating at a speed of one thousand miles per hour!
- The highest temperature in the lower mantle of the earth is about seven thousand degrees Fahrenheit, and the temperatures at the inner core are close to the surface of the sun!
- The earth's magnetic field allows us to be guided by compasses, but birds also use it to navigate because they have a small piece of metal in their beaks.
- If you held a piece of string and someone in Europe held the other end, the string would stretch by about an inch each year. The distance between you increases because of the movement of the continents as they rest on top of shifting tectonic plates.
- The movement of tectonic plates can cause continents to shift, as well as earthquakes, and can also form mountain ranges and volcanos.
- The longest mountain range in the world, the Mid-Atlantic Ridge, is 90 percent underwater.
- Mount Everest is approximately twenty-nine thousand feet tall, which is taller than five thousand adults standing on each other's shoulders!
- Some estimates say the tallest parts of the Appalachian Mountains once towered at thirty thousand feet high—as tall as Mount Everest!
- The country of Panama is a large isthmus connecting South America to the rest of North America.
- Our oceans fill up the largest basins on Earth. Oceanic basins are so large that we can find underwater mountains and valleys within them. The deepest valley on Earth is in the ocean, the Mariana trench, which is seven miles deep! Vast portions of it, and many other areas of the ocean floor, remain unexplored. In fact, some scientists say that we know more about the surface of the moon than we know about Earth's own ocean floor!
- Mammoth Cave National Park in Kentucky holds a cave system that covers over eighty square miles!
- The stratosphere—the second layer of our atmosphere after the troposphere—extends from 7.5 miles high up to around 32 miles above the earth's surface. Think about how high that is, and we are only on the second layer! The next layer is the mesosphere, the top layer of it being 62 miles (or 330,000 feet!) away from the earth's crust. Less is known about this layer, perhaps in part because it is so far from the earth. The last two layers of our atmosphere are the thermosphere and exosphere, which are both *huge*; the thermosphere is more than 300 miles thick, and the exosphere is ten times wider than that!
- Just over 70 percent of the earth is covered by water. Overall, if you took all of the earth's water, from the oceans and lakes and rivers and from the ice in glaciers and even in the atmosphere, and poured it into a gigantic bucket, it would measure about 332 million cubic miles, or quintillions of gallons, which is simply a number of gallons that is so large, it is difficult to imagine. And yet, not all of it is accessible to us or other organisms. For example, we can't use ocean water for drinking because it has too much salt in it, and over 95 percent of Earth's water is in the oceans.
- Even though water really is everywhere on Earth, it is estimated that less than 1 percent is readily accessible, most of it being in oceans, the atmosphere, or in polar ice caps.

- The process of evaporation has important effects in helping to cool down the earth's temperature.
- Plants help with runoff issues from excessive rainfall by soaking up water in their roots and other tissues; they also help remove harmful toxins.
- Meteorology is the study of the processes in the atmosphere, especially those related to important weather events. One important tool that helps them is RADAR (this stands for RAdio Detection And Ranging). Radar detects things in the atmosphere by sending out signals (radio waves) that then deflect or bounce off storm clouds and precipitation. Using this information, meteorologists can build maps that show the current weather conditions across a continent. We then have a big weather map that shows the relationships between temperature, pressure, wind, and other factors, which allows meteorologists to predict upcoming weather.
- There are many exciting jobs in meteorology: chasing storms to better understand them, studying the atmosphere at NASA, or even working for the Department of Defense!
- Thunderstorms, complete with lightning, form when warm air rises quickly. The quick movement actually changes the charge on the air molecules (a charge like the one formed when you rub a balloon on your hair). The top of the thundercloud has a positive electrical charge, and so does the ground, but the bottom is negatively charged. These charged particles attract one another and rush together, forming the bolt of energy we call lightning. The whole process generates a lot of heat, which causes the air nearby to rapidly expand, creating sound waves (thunder!).
- Of all the energy that reaches us from the sun, approximately 25 percent is bounced back to space by the atmosphere and cloud cover, another 20 percent is absorbed by the atmosphere and clouds, and 5 percent is reflected by the surface of the earth, leaving only 50 percent that is absorbed by the earth itself.
- Energy from the sun hits the equator *more directly* than at the poles, so it receives more energy, warming that part of the earth more. The overall result is that average temperatures generally decrease as latitude increases (the farther you move away from the equator, or said another way, the higher or lower you get on the earth).
- Mountains affect rainfall by casting a "rain shadow." Wind traveling across the earth is forced upwards when it reaches mountain ranges, which condenses the moisture and increases rainfall. But this occurs on the side of the mountain (east or west) where the wind blows into the mountains most often (this pattern varies by latitude). The air has lost its moisture already as it passes over the mountain, so the other side of the mountain is drier because of this rain shadow.
- Coastal areas have less variability in temperature than places in the middle of continents. This happens because water (which the ocean, it turns out, has a lot of) absorbs and holds a lot of energy. In hotter months, some of the energy warms the water. In cooler months, the ocean releases that energy and makes for a milder winter. Because of all the water, coastal areas and islands also experience a lot of precipitation. Together, these effects are sometimes called maritime effects, or maritime climate (the Latin *mare* means "sea").
- Lightning strikes can play an important role in the cycling of nitrogen through the environment. When lightning strikes the ground, it "injects" nitrogen into the soil by converting the nitrogen into different chemical forms. Decaying plants and animals can also help put nitrogen into the soul when they decompose.
- There is so much carbon within life on Earth that we sometimes call living organisms "carbon-based lifeforms."

- Climate helps determine what plants are found in a given region, and plants help determine what kind of animals are found there; taken together, this determines what kind of biome you would find there.
- Not all deserts are hot, as we typically assume. Some are cold in the winter because they are found on high plateaus where the air is colder, and even other deserts can become cold at night; this is because deserts are so dry—they have low humidity—that when the sun no longer heats the desert, the heat from the day doesn't stay trapped at night. Some deserts can get down to below forty degrees Fahrenheit at night.
- If you know how much rainfall a given region receives, and what the average temperatures are, you can predict what kind of biome will be found there.
- There are some parts of the earth, near the equator, that essentially have no seasons like we are used to, staying within the same temperature range all year long.
- Due to dry and rainy seasons found in some biomes, animals must migrate and "follow the rain" to stay alive. More than a million wildebeest, for example, migrate hundreds of miles each year following the rains.
- Most temperate grasslands on Earth have been converted for use in agriculture or raising livestock, in both the United States and elsewhere, but you can visit the Tallgrass Prairie National Reserve in Kansas to see an example of the tallgrass prairie that once covered much of the center of North America.
- Some animals may migrate thousands of miles to leave boreal forests, or any biome with snowy winters, to seek out warmer places in the winter.
- The warmer a given region is, the more diversity of plant life there will be.
- Some mammals, like snowshoe hares and least weasels, replace their brown fur with white during snowy winters to match the background. This camouflage technique protects them from predators who go out searching for them.
- In the tundra, the top layer of the soil freezes in the winter, which makes life difficult for plants. Underground, a layer of soil stays permanently frozen; this is called permafrost. The top layer thaws in summer months, creating temporary ponds and pools of water that drain slowly, since the frozen permafrost layer helps prevent the water from soaking deep into the ground. This creates great conditions for insects like mosquitoes, which then attracts birds who like to eat them.
- Mosquitoes survive winter conditions by having compounds in their blood that act as antifreeze to prevent them from freezing solid during the winter.
- Trees that grow in regions where fires are common have thicker bark to protect them from the flames, and other plants may have deeper roots to protect them as well.
- The Atacama Desert in South America is likely the driest place on Earth, such that some locations in the desert have not had measurable rainfall in decades! How do plants and animals survive in such a place? Cactuses are the classic desert plant—whether the cardón of the Atacama or the saguaro of the Sonoran Desert. They have traits to retain water (thick, fleshy stem) and reduce water loss (needle leaves with a waxy outer layer) to survive between rainfall events. Other plants simply wait for rain as seeds, only growing when rare rains arrive. This means the desert is a dramatically different and beautiful place after it receives rainfall.
- Many desert animals are nocturnal—meaning they are active at night—like bats, tarantulas, ringtail cats, and so many more. This helps them avoid the heat of the day.

KEY TERMS

Abiotic factors – *Chapter 2*: A nonliving condition or thing, such as climate or the movement of tectonic plates, that influences or affects an ecosystem and the organisms in it.

Atmosphere – *Chapter 4*: The term used to describe all of the gases (oxygen, carbon dioxide, nitrogen, and others) that surround a planet.

Atmospheric circulation cells – *Chapter 7*: The cycle and movement of air around the earth, which affects annual rainfall amounts, seasonal variation in rain, and other parts of the climate. Overall, there are three major atmospheric circulation cells in each hemisphere, formed at certain latitudes. These cells help break up the globe into three different regions: the *tropical zone* (from 30° N to 30° S), the *temperate zone* (a band around the earth from 30° to 60°, both in the north and south) and *polar zones* (above 60° latitude, north and south).

Atoms – *Chapter 8*: Tiny particles of matter.

Barometer – *Chapter 4*: A tool used to measure atmospheric pressure.

Basin – *Chapter 3*: A dip or depression in the landscape, sort of like a bowl (though they do not need to be round). Erosion can slowly form basins over long periods, or shifts in tectonics and earthquakes can cause them to form more quickly. Some basins fill up with water to form lakes and even oceans.

Biodiverse – *Chapter 10*: A habitat on Earth with many different kinds of animal and plant species.

Biodiversity hotspots – *Chapter 10*: A region on Earth with an especially high count of unique species.

Biome – *Chapter 9*: A distinct collection of plants and animals, or biological community, found within a particular habitat type (like a desert or a rainforest).

Biosphere – *Chapter 8*: The region of the earth that consists of all the living organisms (bacteria, plants, animals, etc.) located between the atmosphere and the lithosphere (crust and upper mantle).

Boreal forest (or taiga) – *Chapters 9, 10, & 11*: A biome with vegetation composed primarily of cone-bearing needle-leaved or scale-leaved evergreen trees, found in northern circumpolar forested regions characterized by long winters and moderate to high annual precipitation.

Buttes (mesas) – *Chapter 3*: Isolated hills with steep, cliff-like sides and a flat top. These terms are similar, though some say a butte is higher than it is wide and a mesa is wider than it is high. Either can be formed as erosion and weathering act on a plateau and leave an isolated "island" of rock on its own.

Carbon – *Chapter 8*: An important element to biological organisms because many molecules in living things contain carbon (including sugars and other carbohydrates, fats, proteins, and many components of cells and tissues); there is so much carbon that life on Earth is said to be made of "carbon-based lifeforms." It cycles between the atmosphere, biosphere, and lithosphere in a variety of ways.

Caves – *Chapter 3*: Underground chambers in the earth, formed from rock types that dissolve in water and acid (like limestone). Slightly acidic water seeps underground and slowly dissolves the rock to form caves.

Chaparral (or Mediterranean scrubland) – *Chapters 9 & 11*: A type of biome; the dominant plant type here varies between trees, scrubs (bushes and shrubs), and grasses. This biome is seasonally dry and warm, typically found along coastal areas around the Mediterranean Sea or in Northern California.

Climate – *Chapter 6*: The average weather in a place *over long periods of time*; a pattern of weather.

Cloud – *Chapter 4*: A visible mass of condensed water vapor and other small particles floating in the atmosphere, usually high above the ground. Clouds can appear in many forms, including *stratus* (low to the ground, formless and flat), *cumulus* (distinct edges, puffy and cotton-like), *cumulonimbus* (cumulus that develop into large, towering clouds that accompany storms), and *cirrus* (very high in the sky, thin and wispy).

Collection – *Chapter 5*: The process by which bodies of water "collect" runaway water that comes about as a result of precipitation (more likely to happen with an excess of precipitation).

Condensation – *Chapter* 5: The conversion of water vapor back into water once the vapor cools down; this process is important for the formation of clouds.

Coniferous trees – *Chapter 10*: Found in many places on Earth but are the prominent plant species in temperate rainforests; often referred to as evergreens because they keep their green leaves throughout the year.

Continental effects – *Chapter 7*: The effects on a climate in a region that is away from the ocean, in the center of a continent.

Convection – *Chapter 7*: The transfer of heat through air molecules.

Core – *Chapter 1*: The inner most layer of the earth, accounting for about 15 percent of the earth's volume, made up of an outer and inner layer. The outer core is liquid metals, mostly iron and some nickel, sulfur, and other elements (the movement of these metals give the earth its magnetic field). The inner core of the earth is a solid ball of iron, with a smaller amount of nickel and things mixed in. This solid metal ball at the center is about 1,500 miles across. The temperatures at the inner core may reach close to temperatures at the surface of the sun.

Crust – *Chapter 1*: The outer most layer of the earth, accounting for about 1 percent, consisting of igneous, sedimentary, and metamorphic rocks.

Deciduous – *Chapter 10*: Plants that lose their leaves and regrow them each year.

Denitrifying bacteria – *Chapter 8*: Bacteria that use up molecules with nitrogen from the soil and then release the nitrogen as a gas back into the atmosphere.

Density – *Chapter 6*: A measurement of how many molecules are packed into one area; more molecules means higher density.

Deposition – *Chapter 3*: The building up of landforms over time through erosion moving bits of rock, soil, sand, etc.

Desert – *Chapters 9 & 11*: A type of biome; dry areas with only small amounts of precipitation per year. Cactuses and other specialized plants dominate these regions, but there are fewer plants in deserts than other biomes. There is some variation in temperatures, as not all deserts are hot, as we typically assume.

Element – *Chapter 8*: A substance that is made up of only one type of atom.

Endemic species – *Chapters 2 & 11*: A plant or animal that is native and restricted to a certain habitat.

Epiphytes – *Chapter 10*: Small plants that grow on other plants in the little bits of soil that collect in the crooks and crevices in trees.

Equator – *Chapter 1*: The imaginary line that runs around the earth like a belt dividing it between the Northern and Southern Hemispheres.

Erosion – *Chapters 3 & 5*: The process by which wind and water break up small pieces of rock or remove soil, detritus, or dissolved materials and deposit those particles elsewhere. It can occur through physical forces (*mechanical erosion*) or chemical ones (*chemical erosion*).

Eutrophication – *Chapter 8*: Excessive richness of nutrients in a lake or other body of water, frequently due to runoff from the land, which causes a dense growth of plant life and death of animal life from lack of oxygen.

Evaporation – *Chapter* 5: When water is heated enough by the sun's energy that it is converted into gas, or water vapor.

Exosphere – *Chapter 4*: The fifth and final (highest) layer of the earth's atmosphere, over three thousand miles thick; where most satellites orbit the earth, so high that the air molecules cannot transmit any sound waves.

Forest – *Chapter 9*: A type of biome; any area where the dominant plant type is trees, which provides lots of shade. Forests are subdivided into *boreal forest* (or taiga), *temperate deciduous forest, temperate rainforest,* and *tropical rainforest.*

Freezing rain – *Chapter 6*: A type of precipitation usually formed when snow passes through a layer of warm air (melting it) but ground temperatures are cold enough that it refreezes when it lands.

Geologist – *Chapter 1*: A person who studies the rocks and other elements and materials that make up our planet.

Gorges (canyons) – *Chapter 3*: Deep, narrow valleys that water carved into rock over the course of many centuries.

Grasslands – *Chapters 9, 10, & 11*: A type of biome; grasses dominate these regions (like an open prairie in the Midwestern United States). Grasslands are subdivided into *tropical grasslands* (also called savannas) and *temperate grasslands.*

Greenhouse effect – *Chapter 4*: The phenomenon by which the earth's atmosphere, through greenhouse gases like carbon dioxide, methane, water vapor, and other gases, traps a portion of the heat it receives from the sun.

Habitat loss – *Chapter 12*: The loss of biomes due to the land being used for logging, mining, agriculture, or any other human activity.

Hail – *Chapter 6*: A type of precipitation when drops of water freeze together high in the atmosphere; larger hailstones are formed when strong winds send them back upwards for more ice layers.

Hemisphere – *Chapter 1*: The two parts, or halves, of the earth, northern and southern, divided by the equator; additionally, there are eastern and western hemispheres divided by the prime meridian.

Hibernation – *Chapter 11*: A long period in a sleep-like or inactive state; some animals hibernate in winter to save energy from a lack of food available.

Humidity – *Chapter 6*: A measure of the water vapor in the air; higher humidity means there is more water vapor in the air. This is reported as a percentage because a measure of 100 percent is all the water the air can hold. This translates to mean that 30 percent humidity feels less wet and sticky than 85 percent humidity.

Hurricanes – *Chapter 6*: A destructive, extreme weather storm involving a rapid vortex of wind; formed from low-pressure areas over warm ocean water. Air moves toward the rising warm air and the low-pressure area. Over time, this process can generate extremely strong winds, often accompanied by rain and flooding. They are longer-lasting storms than tornados.

Igneous rocks – *Chapter 1*: The primary substance that makes up the earth's crust (approximately 90 percent), formed from other rocks that have melted and then cooled in the interior of the earth. Examples include granite, basalt, obsidian, and pumice.

Island – *Chapter 3*: A piece of land surrounded by water on all sides.

Isthmus – *Chapter 3*: A narrow strip of land that connects two larger areas of land and separates two bodies of water.

Landform – *Chapter 3*: A natural feature of the earth's surface.

Latitude – *Chapter 3*: Along with *longitude*, a series of gridded lines that stretch across the earth and allow us to measure and locate various places through a series of coordinates. Latitude lines run east and west (horizontal) and therefore measure how far north or south a location is in relation to the *equator*.

Lava – *Chapter 2*: Magma, or molten rock, that emerges onto the surface of the earth when a fissure or vent opens in the crust.

Lightning – *Chapter 6*: A bolt of energy, an electric current, formed when the quick movement of warm air changes the charge on air molecules, a charge like the one formed when you rub a balloon on your hair; the top of a thundercloud has a positive electrical charge, as does the ground, but the bottom of the cloud has a negative charge. These charged particles attract one another and rush together, forming a bolt of lightning.

Lithosphere – *Chapter 1*: The crust and upper mantle of the earth.

Longitude – Chapter 3: Along with *latitude*, a series of gridded lines that stretch across the earth and allow us to measure and locate various places through a series of coordinates. Longitude lines run north and south (vertical) and therefore measure how far east or west a location is in relation to the *prime meridian*.

Magma – *Chapters 1 & 2*: Hot, semi-liquid, melted rock found in the earth's mantle; when it reaches the earth's surface, we call it lava.

Magnetic field – *Chapter 1*: Caused by the movement of liquid metals in the earth's outer core, this invisible force extends into outer space, surrounding the earth and serving many important functions, including, among others, protecting the earth from radiation. The magnetic field is what allows us to be guided by compasses.

Mantle – *Chapter 1*: The middle layer of the earth between the crust and core, a shell of rock subdivided into an upper and lower part, which constitutes most of the earth's volume (84 percent). Parts of the mantle contain magma, or semi-liquid, melted rock, as a result of the heat generated in the earth's core.

Maritime effects – *Chapter 7*: The effects on the climate caused by the sea, seen the most in coastal areas (the Latin *mare* means "sea").

Mesosphere – *Chapter 4*: The third layer of the earth's atmosphere, about sixty-two miles (or 330,000 feet!) away from the earth's crust.

Metamorphic rocks – *Chapter 1*: One of the substances that makes up the earth's crust, a combination of igneous and sedimentary rocks that have been changed (*metamorphosis* means "change") into a new type of rock by intense pressure. Marble, for example, is a metamorphic rock formed from limestone, a sedimentary rock.

Meteorology – *Chapter 6*: The study of the processes in the atmosphere, especially those related to important weather events.

Microclimate – *Chapter 7*: The climate of a very small area.

Microorganisms – *Chapter 8*: Small organisms.

Neonicotinoids – *Chapter 12*: Chemicals used to ward pests off crops, but which can have a harmful effect on other insects, such as bees.

Nitrogen – *Chapter 8*: An important element to biological organisms because it is necessary to build DNA and protein; it's cycled through the layers of the earth (lithosphere, biosphere, atmosphere) in a variety of ways.

Nitrogen fixation – *Chapter 8*: The process of converting atmospheric nitrogen into the soil.

Nutrient cycling – *Chapter 8*: The cycle of nutrients and elements (nitrogen, carbon, etc.) through the layers of the earth (lithosphere, biosphere, atmosphere).

Orbit – *Chapter 1*: A regular, repeating path of one object in space circling another, as the way the earth revolves around the sun.

Ozone layer – *Chapter 4*: A band of gases found in the stratosphere, approximately between nine and twenty miles away from the earth's surface, where there is a higher concentration of ozone in the air. *Ozone* is a molecule made up of oxygen atoms, and these molecules have the important function of absorbing UV light from the sun; thus, the ozone layer protects us from harmful UV rays and so is sometimes called the *ozone shield*.

Peninsula – *Chapter 3*: A piece of land surrounded on three sides by water.

Permafrost – *Chapter 11*: A layer of soil that stays permanently frozen.

Plateau – *Chapter 3*: A flat, table-like area of land elevated from the area around it on at least one side.

Polar vortex – *Chapter 7*: A strong current of wind in the stratosphere that typically keeps cold, polar air isolated in the polar zone, but when that wind weakens (meaning the air moves less forcefully or quickly), the path meanders southward, letting cold air escape to the south.

Pollution – *Chapter 12*: Anything new or novel introduced into the environment that has harmful effects (can include noise and light pollution).

Precipitation – *Chapter 5*: Once too much water condenses in clouds, more and more water molecules stick together in liquid form, which makes the water droplets heavier, and they fall back to the earth as precipitation; precipitation comes in different forms: rain, snow, hail, or sleet.

Prime meridian – *Chapter 3*: The imaginary line running north and south set near the Royal Observatory in Greenwich, London, England; serves as the start of longitude measurements (i.e., it has a longitude of 0 degrees).

RADAR – *Chapter 6*: Stands for **RA**dio **D**etection **A**nd **R**anging. Radar detects things in the atmosphere by sending out signals (radio waves) that then deflect or bounce off storm clouds and precipitation. Using this information, meteorologists can build maps that show the current weather conditions across a continent. We then have a big weather map that shows the relationships between temperature, pressure, wind, and other factors, which allows meteorologists to predict upcoming weather.

Rain – *Chapter 6*: A type of precipitation when water droplets in clouds become large and heavy enough to fall to the earth.

Rain shadow – *Chapter 7*: The phenomenon of wind traveling across the earth and being forced upwards when it reaches mountain ranges, which condenses the moisture and increases rainfall; this occurs on the side of the mountain where the wind blows into the mountains most often (this pattern varies by latitude). The air has lost its moisture already as it passes over the mountain, so the other side of the mountain is drier because of this "rain shadow."

Runoff – *Chapter 5*: Liquid water that flows over or across land surfaces.

Savannas (tropical grasslands) – *Chapters 9 & 10*: A biome with warm to hot temperatures all year long and a rainy and dry season; grass is the dominate plant in these regions, with fewer trees.

Seasonality – *Chapter 9*: The phenomenon by which a given biome will have different temperatures and amounts of precipitation based on where it is located on the earth.

Sedimentary rocks (sediment) – *Chapter 1*: One of the substances that makes up the earth's crust, consisting of dirt, soil, detritus, and debris. Examples include sandstone, limestone, and shale.

Seismic waves – *Chapter 1*: An elastic wave in the earth produced by an earthquake or other means, which can be measured as it moves through the earth and allow scientists to learn about the makeup of our planet.

Sleet – *Chapter 6*: A type of precipitation usually formed when snow first passes through a layer of warm air (melting it) and then cold air (*refreezing* the water before reaching the ground).

Snow – *Chapter 6*: A type of precipitation that occurs if the air is cold enough; water freezes into ice crystals that grow outwards into flakes.

Strata – *Chapter 3*: A layer or series of rock in the ground.

Stratigraphy – *Chapter 3*: The branch of geology concerned with the study of strata.

Stratosphere – *Chapter 4*: The second layer of the earth's atmosphere, stretching from 7.5 miles up to 32 miles above the earth's service; holds the ozone layer.

Subduction (subducting plate) – *Chapter 2*: One of the types of movement of tectonic plates, when two collide and one goes beneath the other (the one that goes beneath is called the subducting plate).

Tectonic plates – *Chapter 2*: Massive slabs of rock that make up the crust of the earth, fitting together like a jigsaw puzzle. These plates move in a variety of ways, which can cause earthquakes and move continents slowly over time, as well as form mountains and volcanos.

Temperate deciduous forest – *Chapter 10*: A biome more common than rainforests and that receive less rainfall; they have seasonal variation in temperatures and are dominated by deciduous trees.

Temperate grasslands (prairie or steppe) – *Chapter 10*: A biome with temperate climates (warm summers, cool winters) and generally receive less rainfall than forests; the limited precipitation supports grass species better than trees, and so grasses are the dominant plant type. In many parts of the United States, they have been used for agriculture.

Temperate rainforests – *Chapter 10*: A biome with moist conditions because of consistent rainfall, though not as warm as tropical biomes, and unlike tropical regions, possesses distinct seasons. Coniferous trees (evergreens) often dominate temperate rainforests.

Thermosphere – *Chapter 4*: The fourth layer of the earth's atmosphere, more than three hundred miles thick and high enough that there are no clouds and water vapor, and not enough air molecules to transmit sound waves; this is where the International Space Station (ISS) is found, as well as some lower-flying satellites.

Thunder – *Chapter 6*: The booming of sound waves created by rapidly expanding air during the process of lightning.

Tornados – *Chapter 6*: A destructive, extreme weather storm involving a rapid vortex of wind; formed over land when cold, dense air crashes into warm, moist air. The warm air then rushes upward, and this movement can create rotating air movements. The updraft and circular wind pattern can pull more warm air upward, continuing the cycle. The tornado funnel we see is formed by water vapor swirling in the circular wind.

Transpiration – *Chapter 5*: The process of water being released from plants (exits the plant as water vapor).

Tropical rainforests – *Chapters 9 & 10*: A type of biome that lacks seasonality, having high temperatures all year long and heavy rainfall, leading to an abundance of plant life and tall trees and many species of animals.

Troposphere – *Chapter 4*: The first layer of the earth's atmosphere; contains all the air needed for humans and animals to breathe, as well as plants, even if they don't "breathe" exactly like we do.

Tundra – *Chapters 9 & 11*: A type of biome; extremely cold and dry areas, either found at high latitudes (arctic tundra) or high elevations (alpine tundra). Like deserts, fewer plants can grow here than in forests or grasslands. Tundra receives less rain and snowfall than boreal forests.

Ultraviolet (UV) rays – *Chapter 4*: Rays emitted by the sun that can be harmful to our skin (these rays are the reason we wear sunblock while outside on sunny days).

Uptake – *Chapter 5*: The process by which plants take up water and bring it into their bodies through their roots and other tissue; this can help prevent too much runoff, or even increase water quality because some plants remove toxins from the water.

Valley – *Chapter 3*: Low-lying areas running between mountains or hills. Most valleys are formed through erosion caused by moving water, like rivers or streams.

Water Cycle – *Chapter 5*: The many ways in which water moves and changes forms throughout the earth (precipitation, evaporation, etc.).

Weather – *Chapter 6*: Refers to the *current* conditions in the atmosphere at a moment in time, as opposed to climate, which refers to average weather over a period of time (a pattern of weather).

Weathering – *Chapter 3*: The breaking down of mountains and other landforms over long periods of time. Can occur through *mechanical weathering* (when physical forces like wind and water wear down rocks and other forms of land) and chemical weathering (when chemical forces—acidic substances found in nature—slowly dissolve rock).

Wind – *Chapter 6*: The movement of air across the earth's surface.

ANSWER KEY

For Parents

CHAPTER 1

FILL IN THE BLANK

1. **third**
2. **orbits**
3. **spins/rotates**
4. **equator**
5. **crust**
6. **core**
7. **geologist**
8. **magnetic field**

DAYS AND NIGHTS

In the box below, explain why we have day and night.

We have day and night because the earth spins on its axis as it orbits the sun. This means part of the earth is shrouded in darkness (night) while it is spun away from the sun, while the other half is facing the sun (day).

THE FOUR SEASONS

In the box below, explain why we have the four seasons.

We have the four seasons because the earth is tilted on its axis. If your half of the earth is tilted *toward* the sun, it's going to be warmer (summer); if it is tilted *away* from the sun, it's going to be colder (winter). In between these extremes we have autumn and spring.

COMPASS NAVIGATION

In the box below, explain how and why we are able to use a compass to help us navigate. Do outside research if you need to.

If you've ever used a compass, you know the needle always points north, which can help us navigate if we're out in the wilderness. The reason it points north is because the small magnetic pin in the compass is suspended so that it can spin freely inside the casing and be drawn to the earth's magnetism. A compass needle aligns itself and points toward the top of the earth's magnetic field, helping us to orient ourselves and know what direction we're traveling.

MAZE

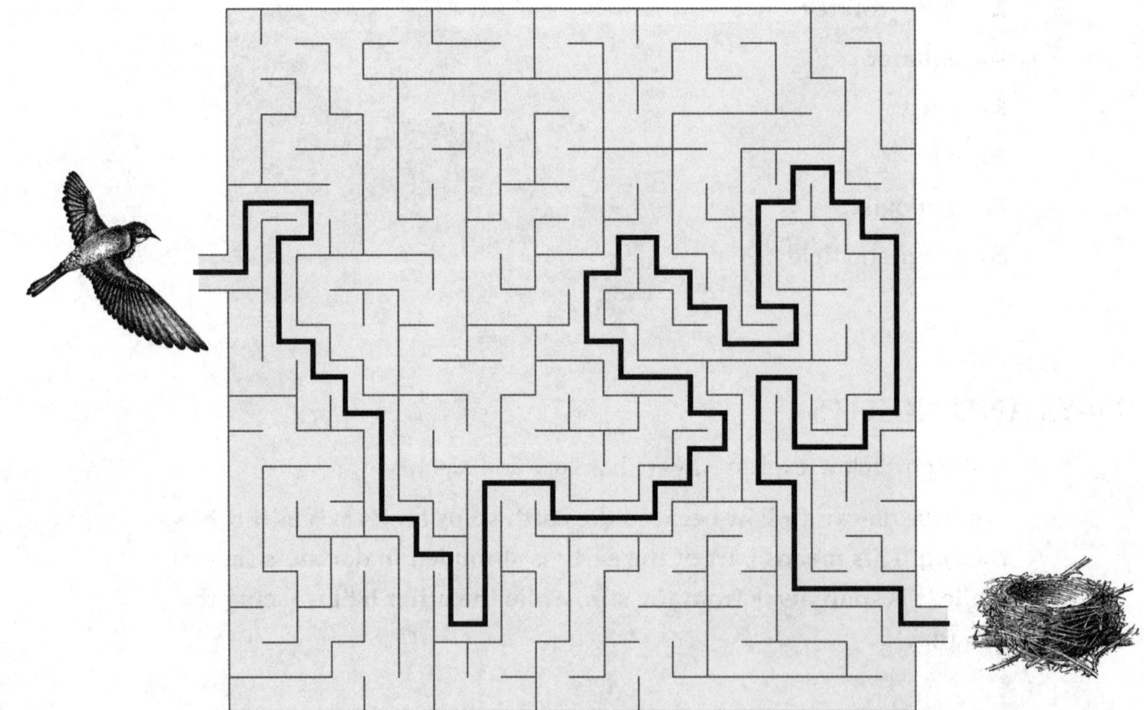

FAITH AND SCIENCE

1. **Pentecost is considered the birth of the Church because this was when the Holy Spirit descended onto Mary and the apostles in the Upper Room, coming as Jesus had promised it would. After this, Peter and the other apostles went out to begin their preaching and baptized thousands. From there, the Church began to grow and spread.**

2. **These nine days of prayer just before Pentecost gave rise to the practice of praying *novenas*, which are a form of worship consisting of special prayers, usually for a given intention, for nine successive days.**

CHAPTER 2

TECTONIC PLATE REVIEW

1. Tectonic plates are:

   Massive slabs of rock that make up the crust of the earth, fitting together like a jigsaw puzzle.

2. The four ways tectonic plates can interact with each other (four ways they can move) are:

   Separate, slide, collide, or subduct.

3. Some results of the movement of tectonic plates are:

   From the four ways tectonic plates interact with one another, the result can be earthquakes, the formation of mountain ranges, volcanos, and valleys, and the separation or connection of landforms.

MOUNTAINS

Facts about mountains include...

- There are only around 1,900 active volcanoes on earth. Some volcanoes eventually become inactive, meaning they no longer release lava or gas from the earth. This in one way that mountains can form.
- Other mountains on earth formed from the pileup of rocks over millennia, or when tectonic plates crashed into each other and the crust buckled upward.

- Mountains can be found on every continent on earth, and even in the ocean. The longest mountain range in the world is the Mid-Atlantic Ridge, which is 90 percent underwater.
- The tallest mountain in the world, if we only count those above sea level, is Mount Everest. It is approximately 29,000 feet tall, which is taller than 5,000 adults standing on each other's shoulders!
- Mountains also have unique effects on living things because higher elevations are colder and have less oxygen (and most animals, like us, need oxygen to breathe). But if heat rises, why are mountaintops cold? The atmosphere is thinner high up in the mountains (fewer molecules of oxygen, carbon dioxide, etc.), and this means less pressure and lower temperatures. Some animals and plants only live on mountaintops because they are adapted to the cooler temperatures. Others only live at the base of the mountains. This means that you can observe different habitats as you drive up a mountain.

CHAPTER 3

UNDERSTANDING THE DIFFERENCE

Weathering and Erosion:

Weathering and erosion are the two main processes that break down mountains and other landforms over time.

Through the process of weathering, larger rocks are broken down into smaller pieces. Water, wind, animals, plants, chemicals (like salts), and extreme temperatures all contribute to the effect of weathering on rocks.

Erosion is defined as processes, like wind and water movement, that break up small pieces of rock, or remove soil, detritus, or dissolved materials, and then deposit those particles elsewhere.

The main difference between the two is that erosion tends to just move particles of land (dirt, sand, rock, etc.) while weathering wears them down completely.

Mechanical erosion / weathering and Chemical erosion / weathering:

There are two types of erosion and weathering, both called mechanical and chemical. Mechanical forces are physical forces, like water and wind, that break down or move particles of the earth, while chemical forces are like acidic substances that dissolve rock and other landforms.

MATCHING

1. G
2. D
3. B
4. I
5. H
6. E
7. C
8. F
9. A

LATITUDE AND LONGITUDE

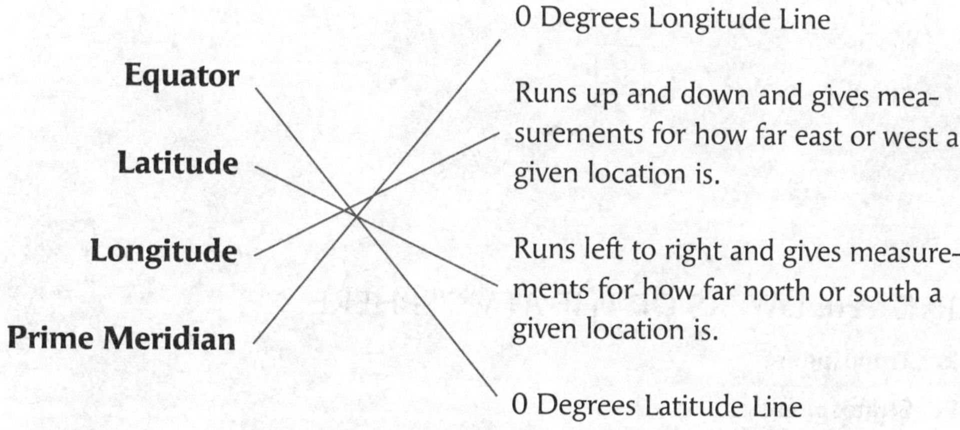

CHAPTER 4

CROSSWORD PUZZLE

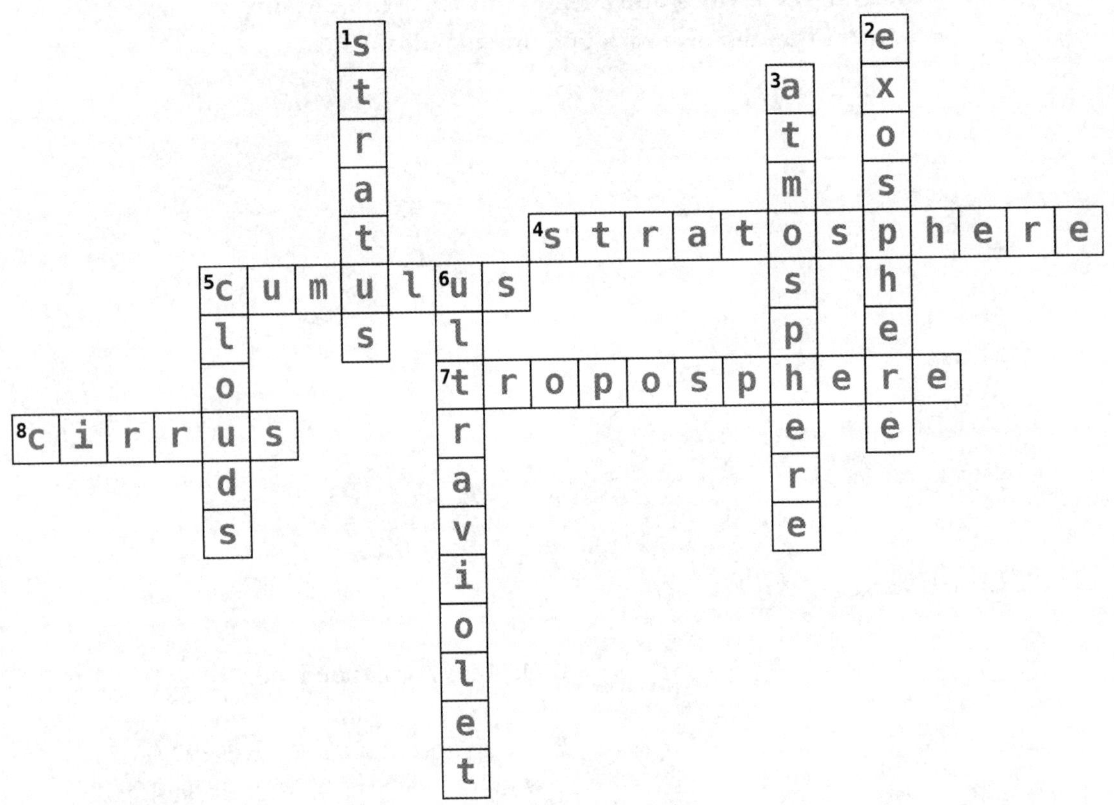

ORDERING THE LAYERS OF THE ATMOSPHERE

1. **Troposphere**
2. **Stratosphere**
3. **Mesosphere**
4. **Thermosphere**
5. **Exosphere**

GREENHOUSE EFFECT

What is the greenhouse effect? Why do we call it this? How is it a good thing but also could be bad if we have too much of it?

Like any object that is warmed up, the earth (being warmed by the sun) slowly releases heat back into space (as infrared radiation). But the atmosphere traps some of this heat. The heat-trapping effect works the same way as a greenhouse does: it lets in light, which warms things up, then traps some heat to keep things warmer for plants to grow.

This is what we call the greenhouse effect. Carbon dioxide, methane, water vapor, and other gases that trap this heat are called greenhouse gases. Without this greenhouse effect, life on earth would not exist in its current form. But it is possible to have *too much* of a good thing—an increase in greenhouse gases has negative effects on the earth's temperature and climate.

TYPES OF CLOUDS

Draw a line from the name for each type of cloud to the picture that represents it.

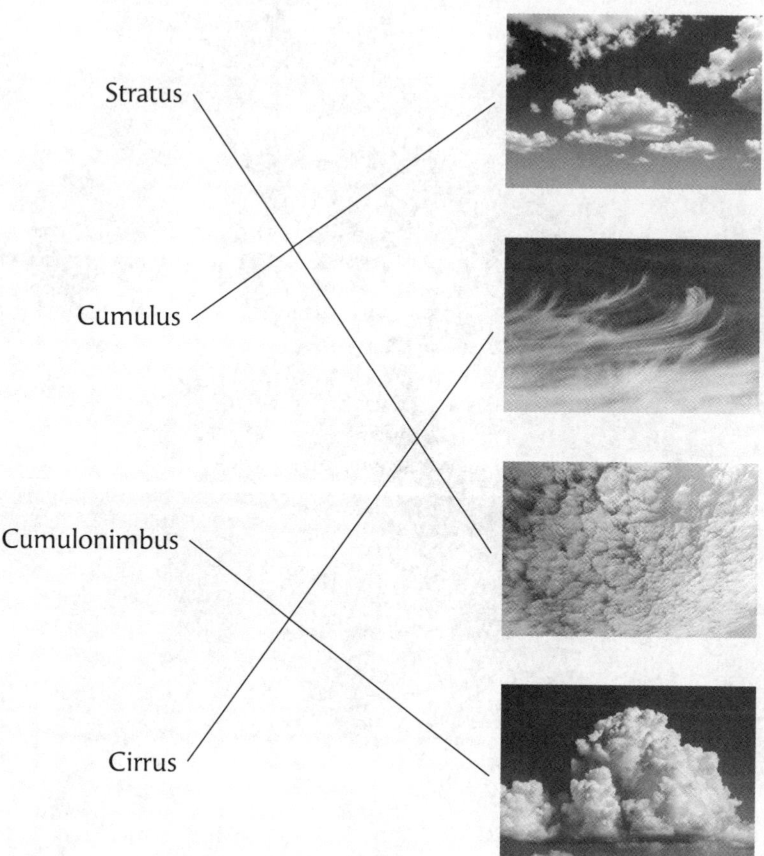

CHAPTER 5

THE WATER CYCLE

In the box below, explain what the water cycle of earth refers to.

Water cycle refers to the many ways in which water moves and changes forms throughout the earth (precipitation, evaporation, etc.). It includes abiotic processes like rainy weather, but it also involves the living organisms that use water, such as plants that take up water from the soil and eventually send it into the atmosphere. It does not refer specifically to rivers and tides and the movement of bodies of water, but those can have an effect on the water cycle.

On each of the blanks, fill in the right term for each stage of the water cycle.

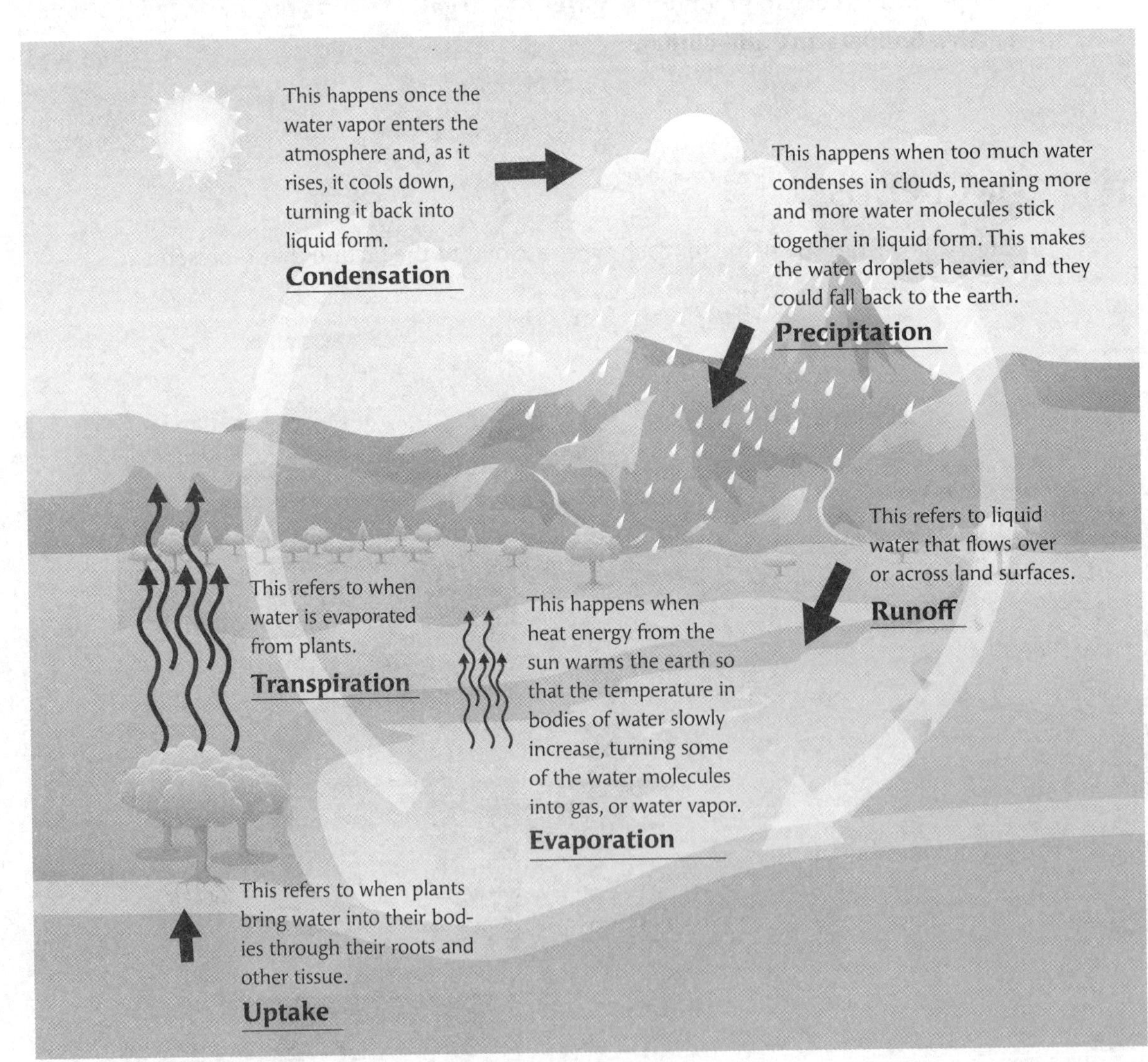

In the box below, or in a discussion with your parents, explain how plants can have positive benefits to the water cycle.

Plants can have positive benefits on the water cycle primarily in three different ways. (1) They can help slow down the flow of runoff as they soak up water, which helps slow down floods. (2) They can help purify water because some plants remove toxins from the water. (3) Through the process of transpiration, they release water vapor into the atmosphere, which eventually returns to the earth through precipitation, thus, giving us water we need to live.

CHAPTER 6

Explain the difference between weather and climate.

Weather is a short-term measurement ("it is raining today"), while climate is a long-term pattern of weather over a period of time ("it is usually warmer in Florida").

List several different kinds of precipitation.

Rain, sleet, snow, hail, freezing rain.

Define meteorology.

Meteorology is the study of the processes in the atmosphere, especially those related to important weather events.

Explain how RADAR works. What does it stand for?

RADAR stands for RAdio Detection And Ranging. It is a tool that helps us detect things in the atmosphere by sending out signals (radio waves) which then deflect or bounce off storm clouds and precipitation. Using this information, meteorologists can build maps that show the current weather conditions across a continent. We then have a big weather map that shows the relationships between temperature, pressure, wind, and other factors, which allows meteorologists to predict upcoming weather.

Why does a clap of thunder follow a bolt of lightning?

The top of the thundercloud has a positive electrical charge, as does the ground, but the bottom of the cloud has a negative charge. These charged particles attract one another and rush together, forming the bolt of energy, the electric current, we call lightning. The whole process generates a lot of heat, which causes the air nearby to rapidly expand, creating sound waves which give the sound of the clap of thunder.

WORD SEARCH

CHAPTER 7

MULTIPLE CHOICE

Chose the best answer.

1. B

2. C

3. A

4. B

5. C

CHAPTER 8

WHAT IS "THE NUTRIENT CYCLE"?

This chapter dealt with the cycle of nutrients through our environment. In the space provided, or in a discussion with your parents, explain what the nutrient cycle is.

The nutrient cycle refers to the movement of nutrients and elements (nitrogen, carbon, etc.) through the layers of the earth (lithosphere, biosphere, atmosphere). This can happen in a variety of ways, including through lightning strikes, through the death and decay of plants and animals, by putting certain fertilizers into the plants that we harvest, and more.

UNDERSTANDING THE THREE SPHERES

1. **Lithosphere: The crust and upper mantle.**

2. **Biosphere: The region of the earth that consists of all the living organisms (bacteria, plants, animals, etc.); located between the atmosphere and the lithosphere.**

3. **Atmosphere: All of the gases (oxygen, carbon dioxide, nitrogen, and others) that surround a planet.**

NITROGEN!

1. What is helpful about nitrogen?

   Nitrogen is important because it helps build our DNA and proteins, which our bodies need.

1. What were some of the ways nitrogen cycles through our environment?

   Nitrogen can cycle through our environment through fertilizers we put in our plants, through lightning strikes, through the food chain (plants and animals dying and decaying into the soil), and through nitrogen-fixing bacteria.

MAZE

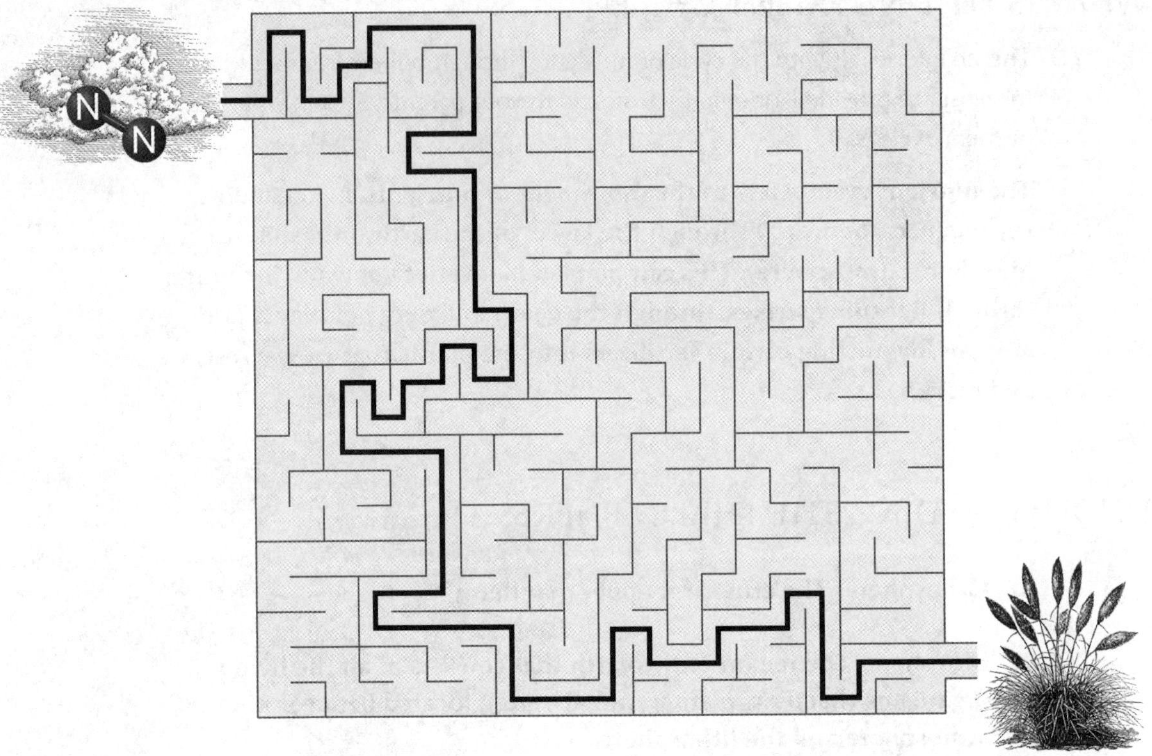

CHAPTER 9

UNDERSTANDING BIOMES

What is a biome, and what are the main factors that influence it?

A biome is a distinct collection of plants and animals, or biological community, found within a particular habitat type (like a desert or a rainforest).

Climate is the main factor that determines the features of a biome, which is influenced by its location on earth. Its climate then determines plant life, which then determines what sorts of animals we find there. All these things, in this order, influence biomes.

NAME THE TERM

What is the term we give to the phenomenon by which a given biome will have different levels of fluctuation (some drastic, some minimal) throughout the year in their temperatures and amounts of precipitation based on where it is located on the earth?

Seasonality

TYPES OF BIOMES

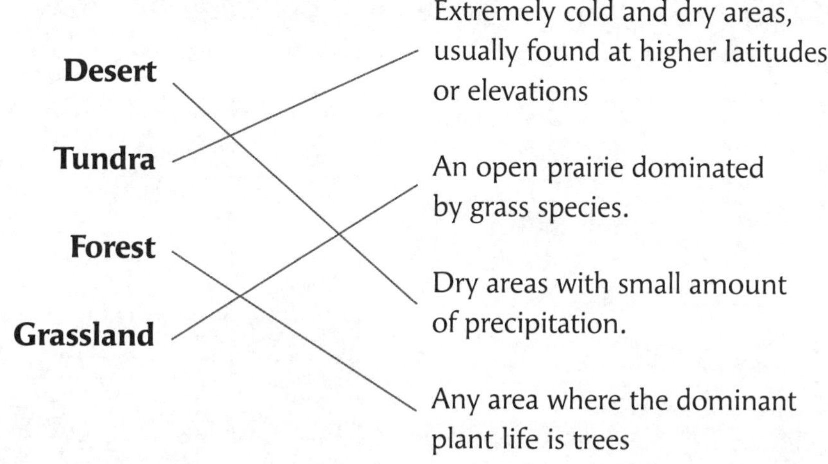

Label the image of each biome below with the correct biome from the list above.

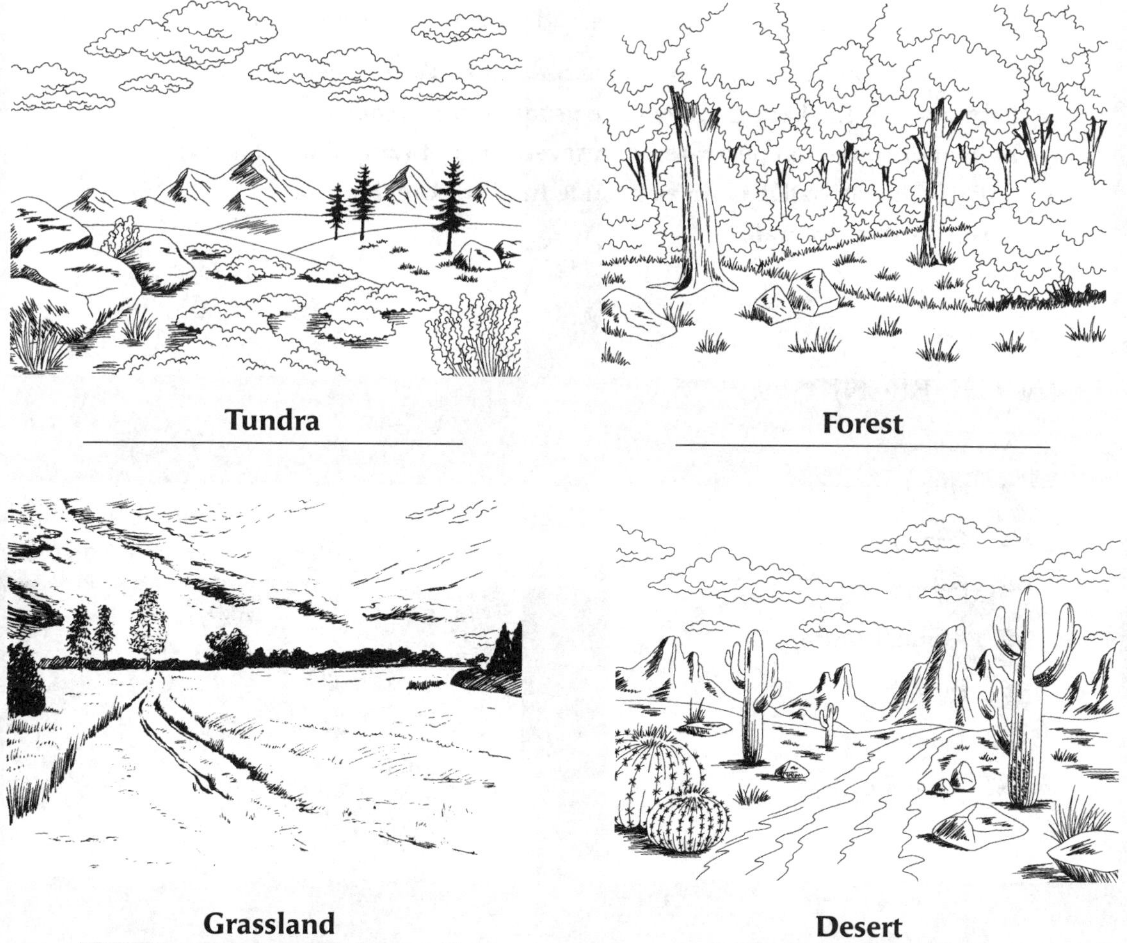

CHAPTER 10

MATCHING

1. D
2. H
3. C
4. B
5. A
6. F
7. E
8. G

CHAPTER 11

EXTREME HABITATS

Why do we call some habitats extreme, and what are some examples?

The word *extreme* means "out of the ordinary." We call certain habitats extreme because they have usual features, such as very high or very low temperatures, or a severe lack of rainfall or a large amount of snowfall. Examples include tundra, boreal forests, deserts, and chaparral.

FILL IN THE BLANK

1. **taiga**
2. **Hibernation**
3. **Chaparral**
4. **Permafrost**
5. **Alpine tundra**
6. **Boreal**
7. **Tundra**
8. **Endemic**

ANIMALS AND PLANTS WITHIN THEIR BIOMES

Cactus: **Desert**

Snow Leopard: **Alpine tundra**

Monkey: **Tropical rainforest**

Palm Tree: **Tropical biomes**

Polar bear: **Tundra or boreal forests**

Black bear: **Deciduous forests**

Fir tree: **Temperate rainforest**

Tiger: **Tropical rainforests / Tropical and temperate grasslands**

Camel: **Desert**

Prairie dogs: **Temperate grasslands**

Oak tree: **Deciduous forest**

Grass: **Grasslands**

Elephant: **Tropical grassland (savanna)**

CHAPTER 12

POLLUTION

Define pollution and give a few examples. Why is it harmful to our planet, and what spiritual lessons can we learn from trying to keep our home clean?

Pollution is anything new or novel introduced into the environment that has harmful effects. Examples include plastic waste in the ocean, or chemicals in the runoff of a factory. Litter and trash from the things we use every day (water bottles, plastic wrappers, etc.) end up in places where they are not meant to be. There is also chemical pollution, such as those we use to keep pests off crops and weeds out of agricultural fields. Still more, there is noise and light pollution, coming from the busy cities in which we live. Artificial light and sound can affect animal life. All these things make our earth less pretty to look at, but can also have harmful effects on animal and plant life, and can even come back to make human beings sick.

The spiritual lesson to learn from pollution is that God gave us the earth as our home, and so we must treat it with respect. If we treat the environment with respect, it is a reflection of how we treat God. Furthermore, practicing the virtue of temperance in the things we consume helps not only the environment but also our souls and interior life.

IMAGE CREDITS

Front cover Skógafoss waterfall in Iceland, Dmitrii Bubonets / The Vajolet towers, Marti Bug Catcher / Trees, nasidastudio / Planet earth, ixpert / Corner ribbon, grebeshkovmaxim / Reaching hands from The Creation of Adam of Michelangelo, Freeda Michaux © Shutterstock.com

pVI Earth © robert_s, Shutterstock.com

p4-5 Earth from outerspace © Motis, Shutterstock.com

p10, 103 Bird © Mur34, Shutterstock.com

p10, 103 Nest © Morphart Creation, Shutterstock.com

p11 Pentecost © Designsoul, Shutterstock.com

p12-13 Geological folds, anticlines and synclines, in Crete, Greece © Nigel Jarvis, Shutterstock.com

p20-21 Hoodoo desert rock formations in San Lorenzo Canyon outside of Socorro, New Mexico © Raisa Nastukova, Shutterstock.com

p28-29 Cloudscape © Chaykovsky Igor, Shutterstock.com

p32, 107 Crossword Puzzle © crosswordlabs.com

p35, 108 Cumulus clouds © Maryia_K, Shutterstock.com

p35, 108 Cirrus cloudss © anmbph, Shutterstock.com

p35, 108 Stratus clouds © SUN IMAGE, Shutterstock.com

p35, 108 Cumulonimbus clouds © Dario Lo Presti, Shutterstock.com

p36-37 Storm over water © Zastolskiy Victor, Shutterstock.com

p40, 109 Diagram showing water cycle © stockshoppe, Shutterstock.com

p41 Floral landscape © Morphart Creation, Shutterstock.com

p44-45 Tropical hurricane eye from space © Viacheslav Lopatin, Shutterstock.com

p52-53 Iceberg © Marti Bug Catcher, Shutterstock.com

p59 St. Kateri Tekakwitha © Nancy Bauer, Shutterstock.com

p60-61 Champferersee lake in the Swiss Alps © Smit, Shutterstock.com

p65, 113 Clouds © sar14ev, Shutterstock.com

p65, 113 Nitrogen, N2 molecule model and chemical formula © Peter Hermes Furian, Shutterstock.com

p65, 113 Plant © Morphart Creation, Shutterstock.com

p66-67 Forest stream © yspbqh14, Shutterstock.com

p71, 114 Tundra © Aluna1, Shutterstock.com

p71, 114 Forest © Aluna1, Shutterstock.com

p71, 114 Grassland © alyona-sergiy, Shutterstock.com

p71, 114 Desert © Aluna1, Shutterstock.com

p74-75 Wild beach of Manzanillo Park in Costa Rica © Simon Dannhauer, Shutterstock.com

p79 National flags © Puwadol Jaturawutthichai, Shutterstock.com

p79 Sacred chalice and divine host © Jorge Ural, Shutterstock.com

p79 Coat of arms of Vatican City State © durantelallera, Shutterstock.com

p79 Celtic Cross © VitAnGen, Shutterstock.com

p79 De la Virgen de Guadalupe © KarminaClemente, Shutterstock.com

p79 Lamb with Cross © AVA Bitter, Shutterstock.com

p79 Our Lady of Perpetual Help © Immaculate, Shutterstock.com

p80-81 Arctic aerial landscape © Andrei Stepanov, Shutterstock.com

p84 Desert landscape © Tartila, Shutterstock.com

p86 Cactus, Bodor Tivadar / Snow leopard, Bodor Tivadar / Elephant, Anastasia Lembrik / Palm tree, Morphart Creation / Monkey, Morphart Creation / Polar bear, Evgeny Turaev © Shutterstock.com

p87 Black bear, Evgeny Turaev / Grass, nikiteev_konstantin / Tiger, Hein Nouwenso / Prairie Dog, andrey oleynik / Spruce pine tree, Bodor Tivadar / Camel, Hein Nouwens / Oak tree, Morphart Creation © Shutterstock.com

p88-89 Floral and mountain landscape © Creative Travel Projects, Shutterstock.com

p94 Third Day of Creation, Genesis, Schnorr von Carolsfeld, Julius (1794-1872) / Lebrecht Authors / Bridgeman Images